Thomas Hunt Morgan

A Contribution to the Embryology and Phylogeny of the

Pycnogonids

Thomas Hunt Morgan

A Contribution to the Embryology and Phylogeny of the Pycnogonids

ISBN/EAN: 9783337407124

Printed in Europe, USA, Canada, Australia, Japan

Cover: Foto ©berggeist007 / pixelio.de

More available books at **www.hansebooks.com**

A Preliminary Note on the Embryology of the Pycnogonids. By T. H. MORGAN.

The position of the Sea Spiders, amongst the Articulates, and the speculations as to the relationship of this to other orders, have been almost entirely based on the anatomy of the adult animal. Kröyer, in 1840, described some embryos, and Dohrn and Hock have carried this a step further. Owing to the extreme difficulties of technique, nothing is known of the internal changes that take place during development.

A study of the germ layers and their subsequent differentiation into organs ought to throw, I thought, some light on the phylogeny of these most interesting animals. Material for work was collected during the summer of 1889 at Wood's Holl, Mass., and I am under very great obligations to Professor McDonald for the opportunity to collect and study at the laboratory of the Fish Commission Station.

Three species of Pycnogonids are found amongst the sea-weeds and hydroids about Wood's Holl, viz.: Phoxichilidium maxillare, Tanystylum orbiculare, and Pallene empusa. During July and August these carry eggs.

The first two genera have a small free-swimming, six-legged larval stage. In Pallene the eggs are much larger, resulting in abbreviated development, and the young leave the cluster of eggs, carried by the male, in a practically adult condition.

In Phoxichilidium and Tanystylum the egg undergoes a regular segmentation into 2, 4, 8, 16; all the segments being equal. After one or two more divisions a condition is reached as shown by Fig. 6. Here the segments have the form of pyramids, with the apices together at the centre of the egg. A nucleus is at the outer part of each pyramid. The egg continues to divide, the pyramids becoming smaller, and then each pyramid divides into an outer and an inner part, each part with a single nucleus. See Fig. 7. Here we have a most perfect delamination, resulting in an outer circle of ectodermal cells and an inner mass of cells. Both inner and outer cells continue to divide. Many of the inner cells now break down, as is seen in the endoderm of many Coelenterate planulae, and the cells of the outer circle become smaller; the line of demarcation between inner and outer cells remains sharp and distinct. It is very difficult to follow out the fate of these two cell masses. Many of the inner cells seem to form a yolk-like substance, with a few scattered nuclei, while the outer cells form undoubtedly the ectoderm of the adult. I believe the endoderm to be formed from some of the nuclei of the inner cells, but I cannot be entirely certain of this, nor have I any observations as to the origin of the mesoderm.

In Pallene, on account of the much larger and more manageable eggs, I have been able to carry out in much more detail the origin and fate of the germ layers. The eggs measure .25 mm. in diameter, and have 125 times the volume of the preceding species. The segmentation is quite interesting, and I had the good fortune to be able to follow it quite far along in the living egg.

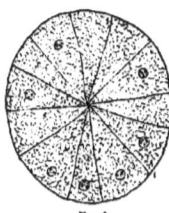 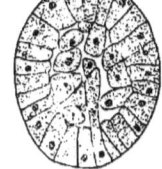

Fig. 6.   Fig. 7.

The first furrow divides the egg into two very unequal parts—a large macromere and a small micromere. The latter is about one-fourth the size of the first. Each segment has a single nucleus. The next furrow—at right angles to last—divides each of the first two into two equal halves. In some cases the larger cell divided five minutes before the smaller, but in other cases the reverse process took place. The furrows dividing the egg into these four parts *nearly* coincide for the micromeres and macromeres. The third furrow divides both micromeres and macromeres into four each, and is at right angles to the first two planes of division. The next furrow is seen to divide the four macromeres into eight, is at right angles to the last two furrows, or is parallel to the first plane of segmentation. At the same time each micromere divides into two, but no definite plane of division is apparent. There are now eight macromeres and eight micromeres. Each of the eight macromeres divides into two in planes at right angles to the last furrow or parallel to the second and third planes of division. This is followed later by a division in the macromere in a plane at right angles to the first and fourth; but only those cells above the horizontal furrow (fourth) were seen to divide. The micromeres were not seen to keep pace with these last divisions, so that the upper (micromere) pole of the egg is covered with a mass of cells of about the same size. Sections of eggs in this stage show pyramidal figures somewhat similar to Fig. 6, but the upper pyramids are smaller, and some do not run to the centre of the egg. Each pyramid has a nucleus in its outer part, and each nucleus is accompanied by a mass of protoplasm which sends out processes into the surrounding yolk of the cell. Soon after this the formation of yolk pyramids ceases, and the nuclei (and their protoplasm) lie at the periphery of the egg. At the upper pole the nuclei are much more numerous, but smaller than at the lower, and the protoplasm forms a thick covering to the egg. Here, also, the blastoderm develops rapidly. At the lower pole there are scattered nuclei at the surface of the yolk.

The early separation of the egg into two unequal parts is apparently closely connected with the more rapid development of the embryo in the region of the smaller segment. About the time when the pyramids become lost (and perhaps at that time) each of the peripheral nuclei divides radially into an outer and an inner nucleus—each of course with its cell protoplasm. It takes place first over the upper pole, and not till very late over the lower area. This is undoubtedly the same thing as the delamination in the smaller eggs. The differences are these—that in Pallene the pyramids do not divide themselves each into two cells, but only the nuclei and protoplasm; and this takes place later at the lower than at the upper pole. The delaminated nuclei remain just under the outer cells—ectoderm—and only exceptionally do one or two wander into the yolk. These nuclei form the endoderm of the mid-gut after having devoured the yolk. The protoplasmic layer at the upper pole becomes wider and the nuclei more numerous, each nucleus being the centre of a distinct cell. At one place may be seen from surface veins an opaque area (much like the early stage of the primitive cumulus in spiders), and sections show that here an invagination of ectoderm is forming—the stomodæum. Around its periphery there is a collection of cells which are, no doubt, the beginnings of the mesoderm. The stomodæum increases in depth, and at this time the

appendages and their nerve ganglia may be seen on the surface. Above
and anterior to the stomodæum are the thickenings of epiblast to form the
brain. On *each side* of the invagination the first pair of appendages arises.
Behind the stomodæum live pairs of large ganglia appear, and on each
side of the three posterior pairs are formed the fourth, fifth and sixth pairs
of appendages.

At the "lower" pole of the embryo (at this time the dorsal and posterior) the nuclei are slowly multiplying and cover the surface with thin
protoplasm. Now the embryo lengthens in the antero-posterior direction,
the appendages and their ganglia become more conspicuous, and there is
seen a slight *invagination in the centre of each ventral ganglion*. There are five
pairs of these invaginations corresponding with the same number of ganglia, and I shall call them the *ventral organs*.

FIG. 8.

Cross sections of embryos show distinctly a wide ingrowth of the surface
epithelium into the centre of each ganglion. The cells of its walls are
rather high, with a clear outer portion, and with large nuclei. Fig. 8
shows such a cross section. These organs close on their outer surface,
and there remains in the centre of each ganglion a cavity, which is rather
longer than wide from side to side, and persists till quite late in embryonic
life.

I need only refer to similar (?) invaginations in Peripatus, and more
especially to a section of a pair of these organs figured by Sedgwick for the
ventral organs of the jaws of Peripatus (*Studies Morph. Lab.*, Cambridge,
Vol. IV, Pt. I, Plate 10, Fig. 4), which easily suggests a comparison with
an *early stage* in the development of the ventral organ of Pallene. Although
I have looked very carefully I have not satisfied myself as yet as to whether
or not there are any invaginations for the brain of the Sea Spiders. The
appendages grow in length, and into each there is pushed an outgrowth
from the mesenteron. These outgrowths contain yolk, and this is covered
by a layer of endodermal cells. Along the ventral half of the embryo there
are scattered mesoderm cells, and these extend into the appendages between
the endoderm and ectoderm cells.

In Pallene the second pair of appendages, which are found in other pycnogonids, never appear, and seem to have been completely dropped from
the ontogeny. The third pair of appendages (the egg carriers of the male)
appear at the time when the young is about to leave the parent. Two pairs
of nerve ganglia develop for these appendages, and each ganglion contains
a ventral organ. Later the two pairs form a single pair with four ventral
organs. The presence of these ventral organs is conclusive proof as to the

erroneousness of Schimkewitsch's hypothesis, that the third pair of appendages are outgrowths of the second pair. The fourth, fifth and sixth pairs
of appendages appear (as already given) simultaneously and quite early.
The seventh appears a little before the embryos leave the parent. Likewise do the ganglia of the seventh appear quite late. There is in addition one pair of ganglia for the rudimentary abdomen. The yolk mass
becomes smaller as it is eaten by the endoderm cells, and exceptionally a
wandering endoderm cell may be found in the mass of yolk. As the yolk
disappears a number of schizocoels appear in the mesoderm between the
body walls and the digestive tract. Dorsally the heart appears as a simple
tube. The stomodæum communicates with the mesenteron and the proc-
todæum forms very late—at the time when the seventh pair of appendages
appear. Soon the embryo leaves the parent and no doubt crawls off among
the sea weeds and hydroids to shift for itself.

The Phylogeny of the Pycnogonids.

It will be impossible to give here the bearing of these embryological facts upon the phylogeny of the group, and I reserve for the future a fuller discussion.

It seems to me, however, that when all the embryological phenomena are taken together they give quite strong evidence for the relationship of the Pycnogonids to the Arachnids. Dohrn and Hoek have each recently reached independently the belief that the group must be considered an isolated one, with a more or less independent origin from the Annelids. I hesitate before offering an opinion against those who are so well qualified to speak authoritatively on the subject. On the other hand their opinions are based largely on the adult anatomy of the group, as little or nothing has been known concerning the germ layers, &c., of these animals; and it is chiefly on embryological grounds that I believe a comparison with the other groups of Anthropods must be based.

It is generally believed that the adults are in many respects degenerate and adapted to a very special habitat—the abdomen has become lost, or almost so, and all traces of respiratory organs are gone, the general surface of the body functioning as such: also that the group is an old one, and not derivable from any existing groups of Arthropods. So far we are together. Without going into details, it does not seem probable that the group is closely related to the Crustacea, nor very closely to the Insects. Here I can only use the *tout ensemble* of the above facts as evidence for this statement. We are then left to decide between an independent origin for the group and an alliance with the Arachnids. If there are any special reasons for an alliance with the Arachnids, I believe such facts must turn the greater weight of evidence towards such a relationship. Briefly then in this connection these considerations must be given:

1. The process of *multipolar delamination to form the endoderm* is, I believe, common to the two groups. We have it represented in its greatest simplicity in the majority of the Pycnogonids, while Pallene furnishes an analogy to the changes which an accumulation of food yolk will cause in this process, and renders a comparison with the Arachnids quite possible. I will refer to Metchnikoff's figures for Chelifer, and to Balfour's embryology for the Spiders (Vol. I, page 119, Sec. Ed.). Here we read: "It appears to me probable that at the time when the superficial layer of protoplasm is segmented off from the yolk below, the nuclei undergo division, and that a nucleus with surrounding protoplasm is left with each yolk column." Compare Fig. 6 and 7, and see account of Pallene.

2. The formation of an opaque area (Pallene) at the place where the stomodæal invagination appears.

3. The early formation of *mesoderm at this place*—the primitive cumulus of Spiders. (?)

4. The general mode of appearing of ganglia and appendages.

5. The body cavity of the appendages and the early presence of mesoderm.

6. The formation of *endodermal pouches* from the mid-gut into the appendages, these pouches *containing yolk in the embryo*. Compare Chelifer and Spiders.

7. The large "upper lip" of Chelifer suggests an homology with the proboscis of pycnogonids.

8. The first (Chelate) appendages appear at the sides of the stomodæum and subsequently move forward, and are innervated from part of the supra-oesophageal ganglia (brain). They will in this bear out a close comparison with Chelifer (or with Arachnids.)

9. The lumen of the invagination of the stomodæum is triangular in

outline and remains so in the adult. Schimkewitsch describes a similar triangular invagination in the Spiders, and compares it directly to that of the Pycnogonids.

The full meaning of the ventral organs I cannot discuss now. I have compared them to similar organs of Peripatus. It may be that in this respect the Pycnogonids show a very primitive structure, common to them and to Peripatus, and if so, traces ought to occur most probably in other Arachnids.

The absence of brain invaginations would be a more weighty objection against the relationship of the two groups, and really the only good objection I know from the embryology.

The openings of the reproductive organs of the adult on the legs cannot be fairly urged against my comparison, for we have so far no explanation of the meaning; and on the other hand this gives little better foundation for a relationship with the Annelids.

All the above comparisons are not of equal weight, and some may be wrong; but taken all in all, I must appeal to them to bear out the hypothesis of the relationship of the Pycnogonids to the Arachnids.

Baltimore, March 15, 1890.

ς

A Contribution to the Embryology
and Phylogeny of the Pycnogonids.
In the year 1767 Karl Linné in the twelfth
edition of his Systema Naturae described under
the name of Phalangium, a Pycnogonid &
here for the first time is the question raised-
whether the group is to ranged under the
Arachnids or the Crustacea.
A hundred years elapsed and the problem
remained unsolved; the group was
placed now here, now there, now amongst
the Crustacea & next amongst the Arachnids.
Then Prof Dohrn attempted the solution
from the standpoint of Embryology,
instituting a comparison 'or even identity'
between the Pycnogonid-larva & the
Nauplius, believing the Pycnogonids
to have diverged from the Crustacea at
this point.

During the following twenty-five years opinions once more vacillated between Arachnidan & Crustacean affinities.

Recently Prof Dohrn & Dr Hoek have each independently monographed the group, placing the morphology of the order on a very firm basis.

Independently likewise they each reached the conclusion that the group is to be placed neither with the Arachnids nor with the Crustacea, & the three groups have come down in parallel lines

The early stages of the embryology of the Sea-spiders have been practically untouched, & before any final descision, as to the affinities of the group is to be made, these stages in the development should be known & take equal rank with Comparative Anatomy in disentangling the affinities of the group.

For many reasons the present paper attempts in no way to give a complete answer from the embryological side. The very great difficulties of technique had slowly to be overcome & the time at command prevented a detailed description of the different organs arising from the germ layers, so that much remains that might be done.

In the Summer of '89 material for work was collected at Woods Holl, Mass. Through the courtesy of Prof. Mac Donald I was enabled to collect & study at the station of the U.S. Fish Commission at the this place. To Prof. Mac Donald I am also indebted for many other kindnesses extended during my stay at Woods Holl. Three genera of Pycnogonids — each with a single species — are to be found at this place, viz. Pallene empusa, Phoxichilidium

maxillare Smith (Cnoplodactylus lentus. Wilson) & Tanystylum orbiculare. During July, August, & September these are found with eggs. Pallene inhabits the hydroids (Tubularia. Pennaria) on the piles of the wharves & is common on the red-sea-weeds below low-tide mark. The hydroids or sea-weeds as soon as collected were brought into the laboratory & worked over piece by piece. Each bunch was in turned swished rapidly backward & forwards in a dish containing a small amount of water so that the Pycnogonids were shaken loose & could be easily picked out. The other genera were more easily found & on separating the masses of hydroids & could be easily seen clinging to the stems. The males of Pallene carry on each pair of ovigerous legs a small

bunch of eggs. Each bunch contains
from one or two to fifteen or twenty eggs
The eggs of Phoxichilidium & Tanystylum
are individually much smaller than the
last, but very numerous so that the bunches
are much larger especially so in the former.
Phoxichilidium carries several bunches strung
along on each ovigerous leg of the male, the
bunches are white and very conspicuous
against the purple color of the adult.
Tanystylum has smaller bunches of eggs
with the individual eggs large, & the
masses are carried so that they form a
circle of clusters held against the ventral
side of the male.
The adults with eggs were put into
alcoholic picro-sulphuric acid for several
hours & then gradually carried through
different grades of alcohol of increasing strength.

Other methods of hardening gave far less satisfactory results, i.e. boiling water or Flemming's solution.

To prepare the eggs & embryos for study they were passed through absolute alcohol (1.hr.), turpentine (2-4.hrs), soft paraffine (1.hr), hard paraffine (1.-2.hrs,) They were cut in paraffine & fixed to the slide with albumen fixative. Then back again through turpentine, absolute alcohol, 95%, 80% 70% alcohols to Kleinenbergs Hæmatoxylin where they remain for a very long time (12-48hrs), then washed fifteen minutes in acid alcohol & up again through the alcohols to turpentine & into balsam. In Pellucae each egg was, in many cases, pricked with a very sharp needle before going into absolute alcohol. It is necessary to do this under a dissecting microscope.

By these methods very excellent results were often obtained & after many failures of other methods was found to be the only satisfactory one.

In Pallene the larger size of the egg makes a study of the earlier stages much easier but the other genera have a much simpler developement & it seems better to give first an account of these.

To Prof W. K Brooks I am greatly indebted for help & suggestions during the work.

Tanystylum & Phoxichilidium

The eggs of Tanystylum measure .08 mm in diameter (in preserved specimens) & those of Phoxichilidium .05 mm.

In both animals there is a regular segmentation. In Plate III figure a u.

a surface view of an egg of Phosphtulidium, divided into two equal parts; in figure b an egg into four cut with the segments shifted around, & figure c an egg into eight equal parts Similarly figures d & e for the two & four cell stages of Tanystylum The eight celled stage I did not obtain for this species.

Figure f is a surface view of an egg of Tanystylum at about the twenty-four celled stage & an optical section of an egg at this stage shows each cell to run from the periphery to the center of the egg where they all come together at a point Each cell is thus pyramidal in shape & contains a single nucleus. In Plate I figure 8 there is a section of an egg at a later stage than the last. This and the following sections

are from eggs which were cut in paraffine.
The section is a later stage than the
preceeding surface view so that the pyramids
are narrower than before.

Figure 11 shows a somewhat older stage for
Proxenetidæum. Very soon after this
— perhaps after the next cleavage — a
most important change takes place in the
egg. I have not seen the actual change
in the living eggs, but serial sections
leave not the slightest doubt as to the
process. Each nucleus divides radially
into two & this is followed by division
of each pyramid into two parts in a
plane at right angles to the radius,
of the circle, lying in that pyramid.
This is shown by Pl. I figure 9 for
Tamystylum.

Here it is seen that by a process of

multipolar delamination. The egg divides
into two germ-layers — an outer
peripheral circle of cells o an inner
mass of cells; these latter soon round
off o leave no trace of the former
pyramidal arrangement. In figure
9 one cell is still seen running
from the periphery to the center. This
section does not show a nucleus in
the inner part. The inner cell I shall
speak of as the entoblast o the outer
circle as the ectoblast. This figure
-9- is somewhat diagrammatic
in as much as only part of the inner
cells is shown for usually they are
more closely packed together than here
shown: o in this case it is due to a
part of the section having broken away
(redrawn in figure) o set some of the inner cells

In figure 11 is a more accurate drawing of an egg of Timystytum immediately after delamination has taken place. Exactly similar changes were seen in the eggs of Phoxichilidium, but the figures first given serve in every respect for both species. The central as well as the peripheral cells continue to divide but soon the entoblast cells loose their well defined boundaries & the nuclei seem in part to disappear. The result is that we have in the center of each egg, a granular yolk with scattered nuclei in it. Here & there is an indication of a cell boundary. During this time the ectoblast cells have divided tangentially & have become much smaller yet at all times a distinct boundary remains between ectoblast &

ectoblast. Figure 12 for Phoronchiladini.
Ø was an embryo at this stage.

It is here seen that the ectoblast cells
over one hemisphere are somewhat
higher than at the opposite & I find
this very constant in sections of both
species at this stage of developement
After this it becomes difficult to
follow out the fate of the germ-layers.
The outer cells become smaller &
flatter to form the ectoblast & the
inner cells arranged into organs
the most conspicious of which is the
digestive tract.
There is a triangular invagination
to form the stomodaeum & the
proboscis appears between the
first pair of appendages which
have now begun to form.

These appendages are very conspicuous in surface view where they project beyond the surface of the body. Between them appears the slightly projecting proboscis & about the middle of the embryo are seen the second & third pairs of appendages which are small and inconspicuous. Dohrn has given excellent figures of embryos at this stage both in his earlier paper & in his later monograph. Soon after this the egg coverings swell up somewhat & the embryos finally break out of the egg, so that the appendages now can straighten out. In Plate IV. figure IX is a surface view of a larva of Tanystylum. as seen from the ventral surface. Dohrn speaks of this as the Pantopod-larva & Hoek calls it a Protonymphon

Its general characters are shown in the figure. There are three pairs of appendages The first pair chelate with the movable claw working outwards & downwards, The second and third pairs have, distally a sharp spine. in the middle of the limb a large segment & proximally a prolongation of the body which may be regarded as The basal segment of the appendage. Between The first pair of appendages The proboscis projects forwards, with the mouth opening at its distal end. Behind the base of the proboscis is seen the pair of ventral ganglia From there the circum œsophageal commissure ascends upwards around the œsophagus & laterally are given off on each side a pair of nerves T.

The second and third pairs of appendages. Dorsal to these ganglia - four in all, the two pairs being fused, is seen the outline of the mesenteron which ends blindly behind

About the middle of the embryo just back of the pair of ganglia is seen a pair of oblong masses which are the beginnings of another pair of ventral ganglia.

In the basal joint of the first pair of appendages is an opaque mass composed of large cells, from which runs out a duct towards the base of the spine and Dohrn has traced it out to the very tip of the spine, where it opens to the exterior. This organ is a gland and its duct is the long tube in the spine

A dorsal view of the embryo shows the brain, just above or a little in front of the ventral ganglia. On its upper surface are a pair of small pigmented eyes. A transverse section through the body of this Pantopod larva is shown in Plate 1. 13. It passes through the center of the body & cuts below the the large ventral ganglionic mass, & passes through the base the the third pair of appendages. Above it cuts the posterior part of the brain. In the center of the section is the digestive tract. This is cut at a point where it is about to give off diverticula to the first & third pairs of appendages. The first of these is marked D' & the second D".
Below the digestive tract is the second

pair of ganglia which sends out
nerves to the corresponding appendages
Above the digestive tract the section at
B cuts the posterior part of the brain.
The cavity of the body has a few
scattered mesoblast cells in it & a
few bits of broken muscle fibres.
These fibres in the living embryo
served to connect the the mid-gut
to the body-walls.
In the base of the legs are seen
several cells which at M are arranged,
in part, around a central cavity
This cavity does not seem to connect
with the mid gut & it seems very probable that it represents the body cavity
in the legs, that the surrounding cells
are of mesoblastic origin In Patteu
as we shall see later there the

similar cavities which are undoubtedly
mesoblastic in origin.

The mid gut ends blindly behind as
is seen in other sections.

In figure 14 is shown part of a section
through a more anterior part of the body
The section passes through the brain
above & the first pair of ganglia below.
These two are connected by the circum-
oesophageal ring   This commissure
is composed of both cells and fibres
In the next section to the one figured
a broad band of nerve fibres passes
on each side from the brain to the
ventral ganglia & in figure 14 a few
of them fibres may also be seen
In the middle of the commissure lies
the cross section of the oesophagus.
Its lumen is triangular and its walls

composed of a layer of cells which
are clear around the periphery of the
triangle. Around the œsophagus
are a few scattered mesoblast cells.
From the brain B there is part of its
substance which projects ventrically
towards the œsophagus & is quite
conspicuous in sections of the
Pantopod-larva in this region
On each side of the brain are seen
a pair of diverticular of the mid-gut.
These go to the first pair of appendages
& are traceable to D' of figure 13.
The first pair of appendages are
as Dohrn has shown innervated
from the brain.
The section of the embryo following
figure 13 give a cross section of the
pair of simple eyes. A part of

This section is shown by figure 15

In the middle line of this section
above the posterior part of the brain B
the ectoblast is thickened beneath the
cuticle & immediately on each side
of the middle line appear the sacs
of eyes. Each is seen in section to
be composed of three cells – two
clear outer ones with large nuclei &
an inner much pigmented cell
around the eyes the ectoblast is
quite thick & on each side it sinks
down slightly from the surface, suggest-
ing that the eye may be here increased
in size.

In figure 16 is a section through the
third pair of ventral ganglia, which
we saw on the ventral side of the
embryo posterior to the fused pairs of first organs

ganglia   The section shows that
at this place the epiblast is greatly
thickened & the cells columnar with
clear outer portions.   In the ganglion
on the left of the section the cuticle
suddenly dips down into the center
of the ganglion where it becomes
extremely thin   This was not so clearly
seen in the right hand ganglion but
traces of it were seen here also
These structures I shall speak of as
the Ventral Organs & we shall come
across them again in Pallene where
they are worked out in greater detail.

                    Pallene empusa.
There is great variability in the size
of the eggs on the ovigerous legs
of the males   The average size

of those which seem to be normal eggs is .25 mm; but often bunches contain one or more much smaller eggs, some of which seem to begun to develope. These very small eggs are most probably immature & accidental. Although the eggs of Pallene are much larger than in the preceeding animals — having 125 times the volume — yet the yolk is almost transparent so that nuclear divisions of the segmenting egg are easily seen from the surface.

The adult is remarkably transparent & consequently most difficult to see as it rests quietly amongst the hydroids & sea-weeds. It seems that the eggs have adapted themselves for purposes of protection by thus becoming

transparent more or less time necessarily elapses after the animals are collected before they can be examined so that I have been unable to see the extrusion of the polar bodies. The nucleus of the egg & its accompanying protoplasm is extremely large & situated near the center of the egg.

Each division of the nucleus is accompanied by a division of its surrounding protoplasm so that in surface views it is impossible to separate the one from the other. In the figures of Plate III the darker masses in each cell indicate the position of the nucleus & protoplasm but for the sake of brevity I shall speak of this simply as the nucleus.

In the unsegmented egg the nucleus
is seen to elongate & to divide into
two halves & this is immediately followed
by division of the egg itself into two
parts. The first furrow divides
the egg into two unequal segments
Figure A Plate III shows an egg after
the first division. Each segment
contains a single nucleus (Hoek
has made a different observation from
this. see infra). After this
first segmentation the segments flatten
somewhat & remain so during the
resting period. The second division
came in about three quarters of
an hour (to an hour) after the first
The nucleus of the smaller segment
— the micromere — divided first &
then the segment itself into equal parts

This was followed five minutes later
by the division of the nuclei of the
large segment — the macromere — &
then the macromere itself.

In figure B an egg is shown as seen
from above after this second division
which is indicated by 2 – 2

The plane of segmentation of the micromere & macromere are drawn in
this figure as though they coincided.
but this is not the rule. The plane
of segmentation of the micromere may
be turned at as much as 30° (&
exceptionally almost 45°) to that of
the macromeres. I think it should
be considered as one & the same plane
& one & the same division, & that
of individual variations.

The segments again flatten together

& after an interval of about an hour the third division commences. The plane of division is at right angles to the last & is shown by figure C in which the egg is seen from above. Again the plane of micromere & macromere do not quite coincide, the number 2 - 2 indicate the second division & 3 - 3 give the plane of the present (third) division This plane is a zigzag line lying between 3 - 3 If we examine the opposite pole – the lower – of an egg at this stage we find the large macromeres come together as shown in D although there are many differences in this respect. So far the planes of division of micromere & macromere have been supposed to coincide or to be referred to the same

division plane but after this
stage is reached (8 segments) the
planes cannot be considered ident
-cal   About an hour after division
division the fourth rythm comes
on. This division plane is
shown for the macromeres in side
view in figure in E, by the line
4 - 4 and is seen to be parallel to
the first plane of segmentation 1 — 1
The four macromeres become eight.
About the same time – maybe five
minutes before or after – the micromeres
divide. In figure E they are drawn
after division has taken place, and
figure F shows how these cell are
divided — the curved lines indicating
twin-cells   As we see from the very
nature of the division of the macromeres

t is impossible to have their plane
of division to correspond with that
of the micromere and the division
lines of the latter lie on a plane at
right angles to the plane of division
of the micromeres We have now
eight macromeres & eight micromeres
After a resting period of about an hour
the fifth rythm begins.
The nuclei of the macromeres of figure
E show by their elongation how they
are to divide & the plane of division
lies between them or at right angles
to the last division plane
Fig G shows the egg after division
into eight micromeres but the
micromeres had not as yet divided
& remain eight in number Soon
however they do divide but no definite

plane of division for all of the
segments was discovered. There are now
16 macromeres & 16 micromeres.

The next plane of segmentation of the
macromeres — at the eighth rythm
comes on after   another resting
period of an hour. This is shown
by figure H at 6 - 6.——6 - 6. The
two planes are at right angles to the
last two (5 — 5) and are parallel
to the first & fourth (1 - 1 . 4 - 4)
I did not see the sixteen micromeres
divide into thirty-two nor could I
trace them further — there are 32 macromeres & 16 micromeres, or in all 48 cells.
After an hour's interval the macromeres lying between the planes 1 — 1
& 4 — 4 in figure H were each
seen to have two nuclei but this

did not seem to be the case with
the micromeres of the lower pole
The egg had been under observation
for many hours & did not develope
farther so that this last attempt
at division in the upper macromere
may have no special significance
Besides the general facts of segmentation
as just given, several interesting
variations in the method of segmentation of the egg were clearly made out.
In the example given above the
micromeres divided first (at the
second segmentation) giving two micromeres & one macromere. In two other
observations this was also true but
in three other cases the macromere
divided five minutes before the micro
mere. And again in the last case

the two macromeres divided into four before the two micromeres divided into four. is that the greater amount of yolk of the macromere did not seem in itself to retard segmentation. In several later stages (32 micromeres + 16 mi--cromeres) the rythem of macromeres & macromeres given above did not so closely correspond but the micromeres seemed to drop behind.

If the accumulation of yolk has been a very recent event in the egg of Pallene + the variations in size may bear this interpretation — these differences in the method of segmentation may be due in part to the segments of the egg struggling to give synchronous beats at each rythem but being hampered by the presence of the yolk

If the first plane of segmentation of Pallene correspond with the first plane of division of Phocidulidium & Tanystylum — and a priori this seems most probable — is it not conceivable that the acquisition of yolk to one half of the egg might case great changes in the synchronism of segmentation of the two parts; & if so it is easy to imagine we have here an egg in a period of variation — that one or another of the above changes may become fixed for the species?

Later stages of segmentation than those described are difficult to follow or to distinguish between macromere & micromere of the upper pole but in general the cells around the upper pole are

smaller & more numerous than at the lower. Figure H Plate III shows the ventral (or macromere) hemisphere of an egg at a later stage than any of the preceding.

The outer ends of the cells, those seen in the figures are polygonal & the nuclei lie very near to the surface. The whole cell is pyramidal in shape with its apex at the center of the egg with polygonal base at the periphery of the egg, with the nuclei in the base of the pyramids. Each nucleus is still accompanied by a surrounding mass of protoplasm & it is without doubt this protoplasm which is seen from surface views. At this time the cells of the lower pole rarely divide & remain

for a long time almost constant in
size & number, but the cells at
the upper pole undergo rapid changes
& our attention must now be turned
almost entirely to that region.
The next change which we can
see in surface views is at the
upper pole where a whitish opaque
area is forming, & which may be
profitably compared to the similar
formation in Spiders eggs & called there
the Primitive Cumulus. This is
due to an accumulation of cells at
this point where an invagination is
about to take place to form the stomodaeum.   ~~Soon~~ There are
other opaque areas formed at the
surface, which are seen to occupy
definite positions and are the Thickenings

for the brain, ventral ganglia & appendages. These are shown in figures I & II Plate IV. In figure 1 we have a surface view of the head region of a young embryo. In the upper part of the figure are first the two oval thickenings to form the brain or supra œsophageal ganglion. At this stage they just touch in the middle line. Below this and in the middle line is the thickening of the stomodaeal invagination. In the center is a triangular shaped cavity — the cavity of the invagination. The base of the triangle is towards the brain. On each side are the thickenings of the first pair of appendages. In the lower part of these is a slight depression which indicates the line between the claw

& the next segment of the limb against
which it works. Posterior to these various
thickenings appear two rows of opaque
areas, the nerve ganglia
Three pairs of these are seen in
this figure, the first two being somewhat smaller than the third.
On each side of the last pair appear
part of the thickenings of the first pair
of ambulatory legs (the fourth pair
of appendages.
Figure II is a continuation of the
last figure & shows the ventral-
posterior part of the same embryo
The upper pair of ganglia are the second
pair, & were seen in figure I & the
second pair of this is the third of the
last figure. On each side of this third pair
of ganglia are again the first pair of walking legs.

Two more pairs of ganglia follow this third. On each side of the fourth & fifth pairs of ganglia appear the second & third pairs of walking legs. An embryo at a stage later than the last is shown by figure III.

Here the embryo has elongated in an antero-posterior direction so that the figure shows only the ventral side of the animal. Figure IV is a dorsal view of an embryo at the same stage. These figures show that the region about the stomodaeum has grown forward with the opening of the stomodaeum at its distal end. This outgrowth is seen in the figures as a forward prolongation of the embryo in the middle line. On each side of it are seen the first pair of appendages which have

grown forward.

On the ventral side of the embryo
appear on each side of the middle line
five pairs of large ganglia. See figure
III. On each side of the last pairs
pairs appear the three pairs of walking legs, which are longer than
on the last figure, are bent & stand
out from the surface of the embryo.
Posterior to the last pair of ganglia
the embryo has a thickened mass
of undifferentiated substance.

In a dorsal view — figure IV — we
see the brain & behind it the large
yolk mass of the embryo.
The two halves of the brain have
come together & each half is slightly
lobed. Around the periphery of the
yolk appear the six pairs of rudiment

o behind the thickened posterior mass
of the embryo.

Returning to the ventral view of
the embryo — figure III there is seen
a most interesting structure in the center
of each ganglion. Each is an invagination of the surface into a ganglion
o these invaginations are elliptical in
outline with the long axis corresponding
to that of the embryo. These structures
I shall call the Ventral Organs o
by means of serial sections we shall
later study them in more detail
In the next stage, shown by figure V
the embryo is more oval in outline.
The appendages have also elongated o
become more bent. The posterior pair
have begun to grow forward between the
more anterior pairs.

important change is in the fusion of
the first two pairs of ganglia which
now appear as one pair of rather large
ganglia. This pair of ganglia
still shows its double structure by
the presence of two pairs of Ventral
Organs. The third pair of ganglia
also shows a pair of ventral organs
but the more posterior ganglia are
covered by the posterior appendages
so that these ganglia cannot be seen
from the surface.

After this stage is passed the appendages grow enormously in length
& the embryo becomes flattened from
side to side.

A figure of an embryo at this stage
is shown, in side view in figure VI
The first appendage is chelate

front near its base & one also sees
it has moved more dorsally to the
proboscis.

Beyond & beneath this appendage
appears the proboscis which has
much elongated. Near the base of
the appendage is seen part of the brain
This figure also shows   that the yolk
mass is seen to continue into the
center of each ambulatory limb

The three pair of legs are much
bent & each ends in a spine like
process   The body is seen to end
behind in a knob like projection
The ventral ganglia are shown
between the bases of the legs &
the height of each at this stage is
about equal to its length.
In other embryos the yolk

disappear & the sides of the embryo
to become thicker. After this the
embryo lengthens a great deal the
appendages grow much longer &
become segmented. Another
pair of appendages appears behind
the first pair of walking legs & the
knob-like projection at the end of
the embryo is pushed more dorsally
to form the rudimentary abdomen
& one pair of eye spots appear over
the posterior end of the brain.
After these changes have taken
place we reach a stage shown by
figures VII & VIII)
The first figure is a ventral view of
the embryo at a time when it is
ready to leave the parent. The three
pairs of segmented walking legs

become straightened out at the sides of the body. In the figures only the proximal ends are shown

The fourth pair of walking legs appears at the posterior end of the body
The first pair of appendages — the chelicerae are now attached quite dorsally to the proboscis which appears between & below them. Each has three segments including the terminal one which acts together with the second to form the pincers.

On the sides of the body just in front of the first pair of ambulatory legs are a pair of projections, one on each side These are the beginnings of the third pair of limbs the ovigerous legs.
I have seen no traces of the second pair of appendages in the ontogeny of Pallene.

Five pair of large ganglia lie along
the   body & in addition a small
seventh pair
The first pair as we have seen is double
& composed of the first & second pairs
The next pair – the third – is between
the first pair of ambulatory legs, the
The fourth pair of ganglia lies between
between the second pair of legs; the
fifth near the base of the third pair of
legs & the sixth, each of whose ganglia
is smaller than those in front, at the base
of the fourth pair of legs
The small seventh pair of ganglia
belongs to the rudimentary abdomen
& does not lie in a plane with the more
anterior ones but like the sixth is
dorsal to the one in front of it as
partially shown in the figures.

The œsophagus is seen to start at the
distal end of the proboscis & to disappear
in the body where it is not easily followed
on account of its transparency

The prolongations of the mid-gut
into the legs is not shown in the
figures except for the last pair of
immature legs

In figure VIII is a dorsal view of the
same embryo. Four pairs of large
ventral ganglia, and in addition the
small rudimentary ganglia of the abdomen
The fused first & second pairs are
hidden by the brain.

On the surface of the brain are seen
four small pigmented eyes.

The œsophagus is triangular in shape in
cross section with the broad base turned
upwards & this base is similar

surface   At the posterior end of the
animal the rudimentary abdomen
is seen   & at its end is the opening
of the anus

Internal Changes.

Sections of the early stages of segmentation
give little information in addition to
what we see from surface views.
They show that each nucleus is contained in a separate segment, & that
many of these segments run to the
center of the egg.   The cells derived
from the first micromere cannot reach to the
center of the sphere.

Figure 1. Plate 1 is a section of an egg
at this stage when there are thirty-two
micromeres. The upper part of the
figure is probably in the region of the
micromeres   In the center &

appears a clear cavity within the yolk but this is not constant for all eggs. In some it is certainly absent but I am unable to say whether or not it is caused by hardening reagents.

What seems to be a similar cavity is found in the eggs of some spiders. The whole egg is divided up into segments in the form of pyramids.

In some of them in this figure can be traced to the center but other sections show each pyramid of the macromeres to have its apex at or near the center of the egg.

The nuclei are situated around the periphery of the section & the protoplasm which invariably accompanies each nucleus sends out fine pseudopodal filaments into the surrounding

Figure 2 is in the same stage as the last & undoubtedly passes through the micromeres at the upper pole. These micromeres & their nuclei are seen to be smaller than the cells elsewhere & the pyramids are seen to fall short of the center of the egg. There is no central cavity present in this egg.

Sections of eggs somewhat later than the last show that the nuclei of the upper pole have rapidly increased, & that they have migrated to the surface of the yolk which loses its pyramidal structure over the upper surface. The protoplasm surrounding each nucleus fuses with that of surrounding nuclei, though how close such fusion is formed I cannot say.

Such a condition is shown in figure 3 where the protoplasm form a cap over the upper surface of the yolk. The nuclei in this section are abnormally large which is probably due to the hardening reagents, but the section seems normal in other respects. The lower area of the same egg shows three or four scattered nuclei near the surface of the yolk. At this stage of development there is a single layer of cells at the surface of the yolk. Between this stage & the next which I have figured there is a gap. In this next section, figure 4, we see the number of nuclei at the upper pole to be more numerous than before & much smaller. The protoplasmic cap has become larger

& its protoplasm is for the most part without the amoeboid processes of the last figure.

A very important change has taken place between these two stages, viz the formation of an inner layer of cells. Within the area of the cap appear a few somewhat flattened cells in figure 4, & these send out processes into the underlying yolk. Two of these cells are shown in this figure. I have not made out any definite arrangement for these cells but they seem to lie only under the upper surface of the embryo. Where these cells come from & at what time they appear must in part be a matter of conjecture but much light is thrown on their possible origin by a —

study of somewhat older stages.

In figure 5 is a section of such a stage. The number of nuclei in the outer germ layer has nearly doubled and the area itself covers a much greater surface of the egg than in the last stage.

Under the blastoderm are seen five inner cells, with their pseudo-podial extension. The larger number of these cells here than in the last figure is due in part to the greater thickness of the sections. At the lower pole are the same. Then below is the outer germ layer, although they show (as did all the outer at an earlier stage) the protoplasmic extensions into the yolk. This section itself throws little light on the cells of the

inner layer but in other sections at
a similar stage I have found what
seems to furnish the solution of the
problem. From such sections I have
drawn figures 6 & 7

These are taken from the periphery
of the cap of cells at some such point
as P in figure 5. Figure 6 shows
two cells which have just separated
from the outer layer of cells & we
also see beyond them a single
mass of protoplasm with two nuclei
which presumably have just divided.
I have not seen any karyokinetic
figures in nuclei dividing in this
direction & it is possible we have
here a direct division as Heider
has recently shown in Hydrophilus
at a similar stage of developement

although it is equally as possible it
is due to poor preserving-agents
Again in figure 7 we see a single
mass of protoplasm with two nuclei
which have just divided. These I
believe to give a clue to the origin
of the inner cells of the preceding
figures & to point out that they too
have had a similar origin from the
outer layer

Keeping before us the process of delam-
-ination in Phoxorchilodium &c I
think we may regard these inner
nuclei of Pallene to have come from
the outer cells by delamination & even
that we may push the comparison a
step farther & consider that each cell
of the outer layer to have given rise
at one time in its history to an inner cell

o then that the outer cells continued to divide tangentically to form the blastoderm. The reasons for such a belief are there. In cross section the number of inner nuclei are slightly in excess of the peripheral nuclei of the lower pole. See figures 4. o 5. As the outer cells of the upper pole were at the beginning more numerous than the peripheral cells of the lower pole we ought to get, if the hypothesis be true, exactly what we do find.

Further, at the periphery of the blastoderm where the inner cells of the lower pole are added on we can always see such a method of multiplication taking place. The differences between the [illegible]

that in Phoronidium are there;
that multipolar determination does
not take place simultaneously in all
the cells at once but in Pallene
slowly progresses as the cap of cells
makes its way to the lower pole
of the egg, incorporating into itself
as it goes the outer cells which
cells as they are added on first give
off an inner cell

After this stage we pass to older
embryos where these two layers
– ectoblast & entoblast – begin to
differentiate into organs.
The first to appear is the stomatodaeum
which results from an invagination
of ectoblast. This is shown by figure
17 Plate II Here it is seen the ectoblast
at one point has pushed inwards

the periphery of the imagination
appear several cells with branching
& anastomosing pseudopodia.
These I believe to be the first
appearance of the mesoblast.

A few of the cells drawn in the
figure belong with no doubt to the
entoblast & at this stage it is difficult to separate the two

The section passes longitudinally
through the embryo; & just medial
to the stomodaeum there is a thickening of the ectoblast to form the first
ventral ganglion. Under the ectoblast
are found much branched entoblast
cells which are still comparatively
few in number. Whether the
circum-stomodaeal mesoblast comes
from ectoblast or entoblast I am unable

to say. Dorsally to the epiblast the epi-
dermis continues for a short distance
+ then becomes exceeding thin.
From this stage we pass to embryos
of the age represented by figures I & II.
Plate IV. The first figure, Iᵃ,
is a cross section through the stoma-
daeum of I (see line 18 of the
figure) The invagination is deeper
than in the last section & its
lumen is closed The ectoblast cells
of its walls are two layer thick.
On each side of this central invag-
-ination are what appears to be
a lateral invagination. These
supposed invaginations have given
me endless trouble & even now I
feel some uncertainty as to my
interpretation of them. On each

side of these is seen the thickened
ectoblast of the first pair of appendages
The lateral ingrowths are I believe
caused by the growth of these appendages which tend to grow outwards
as they increase in size but are
prevented from doing so by the
the egg coverings & the result is that
the ectoblast becomes folded on itself
to make room for the growing appendages. This view is strengthened
by the presence of somewhat similar
invaginations at the side of the
other appendages
This part which is pushed in would
correspond to the dark furrow between
each appendage of figure I & The
cells forming the stomodaeum
The mesoblast around the stomodaeum

& under the appendages has increased
& is now clearly distinguishable from
the inner covering of the ento-blast,
which lies only at the periphery of the
yolk & between it & the mesoblast.
In figure 14 we have a section from
the same series as the last, but more anterior
& through the region of the brain correspond
-ing to line 14 of figure I.   The
section is entirely in front of the
stomodaeum & cuts the true brain
thickenings of the surface view.
Here again the ectoblast is seen to be
distinctly folded & this folding
(imagination!) is directly continuous
with the last.   At first sight this seems
nothing more than a forward continuation
of the last groove but it is not clear why
the folding of the appendage should exist

any influence over that part of the brain. Again it might be interpreted as folding due to the brain but I can see no good reason why such thickenings — not outgrowths — should produce the groove. There is one other possibility, viz, that these may represent ingrowings into the brain itself. Against such a view is the absence in surface views of any such ingrowings, & also that the folds are directly continuous with, & quite similar to, the groove between the olfactory lobes & the appendages; so that I cannot hold this either as a true solution.

Beneath these ingrowths appears the pair of thickenings for the brain which are seen to be continuous across the middle line. On the inner surface of these lobes are seen a few entoblast cells as shown.

The next figure, 20, passes through the middle of the brain, but is farther forward than the last. See line 20 Fig I Plate III which indicates the place of the section. The cells of the ectoblast seem to pass into those of the brain. Sections still farther forwards, there the two lobes are still cut shows the brain thickenings on each side & separated from each other.

We see in surface views of embryos at this stage that the neural ganglia have appeared & figure 21 is a section through two of them,—the first pair. Not only do we see that the ectoblast has greatly thickened in two places on each side the middle line but the outermost cells of each mass show certain peculiarities. They are much elongated at right angles to the surface & their inner points come together at the middle point of the surface & the

ganglion. The left-hand ganglion in the figure is cut through this central point & the ganglion on the right a little to one side. The nuclei lie in the inner parts of the cells & the outer parts of the cells are clear & transparent while their inner ends are more granular. This difference in the parts of the cells is very constant through the later stages.

The cells form a structure which I have called the Ventral Organs.

In figure 22 is a section of a pair of ventral ganglia from an embryo at stage III. Here there is in the middle of each ventral ganglion a wide invagination lined by columnar cells with their clear outer portions turned towards the cavity of the invagination. The nuclei in these cells are larger than those in other parts of the ganglia & are seen quite often in process

of karyokinetic division. The spindles of the dividing nuclei are in some at right angles to the long axis of the cells or in other parallel to this axis.

A narrow connection of small ectoderm seems across the middle line from ganglion to ganglion. The next figure, 23, is from an embryo at about stage V. The outer edges of the invagination have turned in until they have nearly met above the cavity of the invagination. In other respects the section is similar to the last. At the next stage, shown by figure 24, the arching over is completed & fusion has taken place so that there is a cavity in each ganglion. Each cavity does not run through the whole length of each ganglion but exists only in its middle portion. There is no communication between the cavities

of different ganglia. In this figure each ganglion has increased in size & the cells have become more numerous & further the neighboring ganglia are connected by a cross commissure of fibres. Further the nuclei of the outer cells (those lining the cavity) are now more like the nuclei of the ganglion cells. Later stages show that the central cavity of each ganglion disappears although they seem to persist for quite a long after they have been enclosed by the ganglia. That this structure may have functioned as an organ in some ancestral form, I believe possible first because they are distinctly marked differentiated before any invagination takes place, secondly because their cells are histologically distinct from the surrounding ectoblast & from the cells of the ganglion beneath & lastly from

The arrangement of the cells which suggests a sense organ comparable to the simple ectodermal organs of many animals.

It also seems probable that we have exactly similar structures in the Pauto-pod-larva of the other Pycnogonids.

Returning to stage III we have a cross section of the body drawn in figure 25. The section passes through the middle of the body in the plane of a pair of walking legs. The large mass of yolk is still seen filling the upper part of the embryo. Over the ventral surface of the yolk, the entoblast cells form a continuous covering and are indicated by N in the figure. At the base of the appendage the entoblast sends out a prolongation which is filled with yolk while the entoblast cells are here higher

+ more closely packed than elsewhere.

Into each of the six walking legs is a similar protrusion of entoblast from the mid-
gut. Beyond these diverticula there is
in each leg a definitely marked cavity
in the mesoblast. In figure 25 the
surround mesoblast is shown by M.
As the prolongations of the midgut push
into the legs, their cavities cannot be
made out & have either become lost or are
to complicated to follow out.

Besides these cavities which would seem
to be body-cavities we shall come
to others later in development which seem
to be schizocoels. The surrounding mesoderm
of these is irregular & quite different
from that around the body-cavity proper.
The ectoderm over the appendages is thicker
than elsewhere. The neural ganglia at

this stage have in most cases the ventral organs
although in this figure none are shown
others have closed in their parts.

The ectoblast of the dorsal surface is quite
thin o the mesoblast has not appeared
in that part of the animal

The figure 26 is a drawing of a cross
section through the head of an embryo
at a stage a little later than III but
not so late as V   The section
passes through the brain above o the
second pair of ganglia below. Between
the brain o the ventral ganglia the
stomodaeum is seen. Its lumen
is seen to be triangular in outline
with its base turned upwards.

On each side of the stomodaeum appear
cross sections of the first pair of gut-
diverticula.   A single row of entoblast

cells surrounds the central mass of yolk. Scattered mesoblast cells are found around & between these different structures leaving here & there openings between the cells. Such schizocoeles are shown by SS in the figure.

Going on to stage V we find that sections give little more than was seen in the preceding figures. The prolongations of the mid-gut into the legs have grown longer & as seen from surface views the first & second pairs of ganglia have fused. In stage VI the yolk-mass continues to fill the central & upper part of the embryo, but it now begins to decrease in quantity. This may be due in part to the fact that part of it becomes digested & built up into the tissue of the embryo & in part to its extension into the legs

A cross section of an embryo at a stage a little older than 27, & in a plane of a pair of limbs, is shown in figure 27. The ventral ganglia have increased enormously in size & have now a large amount of 'punct-substance'.

In the middle of the section lies the mid gut with its contained yolk which has decreased very much as compared with preceeding stages.

In figure 28 is drawn a section of another embryo, and at about the same stage as the last, but cuts the embryo between a pair of legs. Above the large ganglion is seen the digestive tract which is completely separated from the body wall by a series of shizo cœls. The uppermost of these, lying in the middle line is the heart, H.

At the sides of the midgut appear
two well marked cavities & a third
between the nerve ganglion the outer
wall. Also below there are two or
more spaces but these latter & perhaps
some of the others are artefacts.

The mesenteron is covered on its outer
side by a distinct layer of mesoblast
cells. Cross section of some of the
legs, eg. in, below the body, are also
shown in the figure.

Between stage VI & stage VII VIII
there is a considerable gap. During
this period the embryo has lengthened
& the fourth pair of walking legs has
appeared. The rudimentary abdomen
has been pushed up dorsally & the
proctodaeum invaginated until at
stage VII it has opened into the mesenteron.

The eye are now seen but at stage
VI they are seen in sections a few
thickenings of the epiblast above the
brain. In stage VII - VIII we find
by sections that the yolk has almost
completely disappeared from the digestive
tract. Cross sections also show the
stings nerve ganglia. The mid-gut with its
diverticula, o on the dorsal part of the
mid gut   The tubular heart, running from
the first walking leg to the third.
Figure 16 is from a cross section through
the posterior part of the embryo a cut
The mid gut at its juncture with the
proctodaeum   Below the digestive
tract D is a pair of small ganglia, V°.
which are the ganglia of the rudimentary
abdomen On account of the shifting of
the abdomen to the dorsal side. These ganglia

have been carried above the last cervical ganglia which are shown at V^b.

But this is not all for this sixth pair of ganglia has itself been affected by the shifting of the abdomen & in turn lies above or back of the fifth pair of ganglia. In the sixth pair of ganglia, as shown in the figure, we have evident traces of the ventral organs. In the upper part of the section & on each side of the middle line are the diverticula of the fourth pair of walking legs. A few scattered mesoderm cells appear in the cavity of the body.

The embryo must leave the mother at about this time for I have never found older embryos attached & only exceptionally ones so old as those last figured.

81

Comparisons.   Comparing the embryology
of the two types represented by Phoronchelidium & Pallene we find that most of
the changes of the latter may be considered
by considering it an abbreviation of the type
represented by Phoronchelidium.

An exact comparison of the segmentation
of the two forms would be of interest but
in order that such a comparison should
be of value the exact orientation of the
segmentation plane would be essential
& such observations are wanting.

It would seem à priori most probable
that the first planes in each must correspond
& that the unequal segmentation of the egg
of Pallene has been caused by the greater
accumulation of yolk in that part of the egg
which corresponds to the macromeres.

It is also probable the smaller segment

82

— The micromeres — correspond either to the
anterior or to the ventral part of the embryo
(which if there is correct it is difficult to
say. The embryo differentiates earlier
in what corresponds to the anterior region
of the adult than over the whole ventral
surface, which suggests that the smaller
cells may have adapted themselves to
this early differentiation; but it seems equally
possible that this differentiation may be due to Phylogenetic
laws in this particular case rather than to
any mechanical connection with the micromere
differentiation. So that for the present the
question must remain unsettled until by
actual experiment (which would not be
difficult) the orientation of the segmentation
planes be determined.
There is an observation on the segmentation of
the egg of Pallene brevirostris by Hoek that

was mentioned in the previous description of
segmentation of Pallene empusa.

He says " Le fractionnement commence
par le fractionnement du noyau, et
seulement après que quatre noyau
sont formés, une premier fractionnement
devise l'oeuf en une partie plus grand
et une autre beaucoup plus petite.
Chaque partie contient deux noyaux
qui dans le plus petit segment sont
plus rapprochés l'une de l'autre que
dans l'autre segment."

I cannot believe this a correct observation
in the light of what I have observed over
& over again in Pallene empusa.
There is a stage which corresponds exactly
with the description but the egg itself
has previously divided into two with a
single nucleus in each segment subsequently

each of these nuclei divides into two
just before the segment itself divides
& at the same time the first furrow
becomes more distinct again as the segments
round off to form the second furrow & this
seems to be the stage which Hoek has
described as the first segmentation.
Nor does it seem probable that the differences
of our observations are due to their having
been made on different species for in
each case the egg is approximately the
same size.
The changes which take place in the formation
of the endoderm of Pallene may also I
believe be referred to the simpler changes
of Phoxichilidium &c and furnish in
this respect an excellent basis for further
comparison with other forms having much
yolk present   In Phoxichilidium

pyramidal segments divide into an outer & an inner cell. while in Pallene the nuclei alone divide & although delamination is still multipolar it is not synchronous over the egg.

A further comparison of the musculature I am unable to give owing to lack of complete observations of the changes of Phoxichilidium. In both types the œsophagus invaginates with a triangular lumen & in each the proctodaeum forms quite late in development.

The few observations I have made on the ventral organs of Tanystylum leave no doubt that it is the same structure as in Pallene. Prof. Sedgwick has described in Peripatus paired ventral organs correspond in number & position with the pairs of ventral ganglia

comparing figure 21 Plate 11 with his
figure for the ventral organs of the jaws
of Peripatus (Studies. Morph. Lab. Camb.
Vol IV. Pt I Plate 10. fig. 4.) a striking
similarity is seen o in addition to this
he says in his account of the ventral organs
that they are slightly invaginated from
the surface.
Whether these structures are in any way
related it is impossible to say but it
is worth while to call attention to the
close similarity both in position o structure
between these organs in the two groups.
A comparison of the appendages o their
order of developement is of interest
Prof Solern has most carefully o fully worked
out the transformations of the larval forms, o
through his skill we have a very thorough
knowledge of the transformation of the embryo.

According to his account the six legged
larvae of the Pygnogonids, with the
Pantopod development passes into the
adult condition by its body elongating
behind the last pair of appendages & the
walking legs appearing from before backwards
in much the same way as Prof Claus
believes the typical nauplius to pass into
the adult

During the growth of the walking legs
the second pair of appendages of the
Pantopod larva has increased a little
in size but the third pair looses its
spines & the whole appendage becomes
a simple prolongation of the embryo
At a later stage when the larva has
increased in size this third pair grows
out again to form the ovigerous legs
In Pallene we have seen that the ....

fifth & sixth pairs of appendages appear
simultaneously in the embryo. The seventh
comes in very much later & the third after
the seventh. The second did not
appear at all in the ontogeny so that
in the young Pallene the only appendages
in the young embryo that correspond to those
of the Pantopod-larva are those of the
first pair.

The development of Pallene has
become so much abbreviated that there
is only a trace of the true Pantopod-
larva found in its ontogeny.

Phylogeny.

There is a general agreement that the Pycnogonids are to be placed within the large group of arthropods, but after this there is the greatest divergence of opinion as to which group of Arthropods they are most nearly allied.

In general there are two prominent categories to which all or nearly all of these theories may be referred.

One lot of workers believe in a Crustacean relationship, & another lot placed the Pycnogonids amongst the Arachnids. It is needless to give here the reasons assigned for these opinions as Prof. Dohrn has given in his monograph of the Pantopoda of the Gulf of Naples a most complete & exhaustive bibliography of the literature

90

The two most important views which we
have at present are those of Dohrn & of
Hoek, but as the latter agrees with
much that the former has given, we may
consider the two together.

Prof Dohrns earlier views of the relationship of the Pantopod larva to the Nauplius
he has completely given up in his more
recent papers.

In an appendix to his paper in the
Archives de Zoologie Experimentale
Dr Hoek gives a summary of Dohrn's
recent theory as to the Phylogenetic
position of the Pycnogonids; & has at
the same time contrasted his own
views with Dohrns so that a translation
of a part of this review will best bring
out the resemblances & differences of
their opinions.

91

" For the Crustacea Dohrn rejects the Nauplius
theory (of Fritz Müller & Claus) & adopts
that of Hatschek, who believes the Crustacea
to have descended from parents which had
the form of Phyllopods, just as these
Phyllopods have in turn descended from
Annelids.   From these same Annelids,
according to Dohrn. The Pycnogonids have
come down. The number of their
segments was originally more numerous
than we now see them, the presence
of a pair of rudimentary ganglia in
connection with the last pair of ventral
ganglia allows us to add one pair ... in
of appendages, & together with the first
all of these appendages were originally
homotypes, ... The received diverticula
from the intestine ... & each appendage
enclosed within itself a reproductive organ.

with a special genital opening.
(The so-called excretory organs of
the palpex & of the ovigerous legs
are the rudimentary sexual organs
of these appendages). The appendages were much shorter than we now
see them, the heart showed many
openings &c &c: The supposed ancestor
that Dohrn reconstructs might be very
well compared to an Annelid
but also notice that Dohrn persists in
his opinion, published in his work
of 1879, that the Pycnogonids have
a parentage neither with the Arachnids
nor with the Crustacea (they have developed by the side of the last & altogether
independently). In this I am
in accordance with Dohrn.
I was struck in the first place by the

"very general presence in the Pycnogonids
of a characteristic larval form (the Protonymphon) & its presence suggested to me
the idea of their descent from an ancestor
resembling somewhat a larva which
took its place by the side of the hypothetical
ancestor of the Crustacea - the Nauplius -
& this by the side of a third (the ancestor
of the Annelids); & that all three groups
would have descended from a common
ancestor. I had tried to obtain in
this way an explanation of the affinities
of the three groups of animals (Annelids,
Pycnogonids, Crustacea).
Dohrn on the contrary took into account
the larval forms but constructed a
Phylogeny comparing together those
animals having a considerable number
of segments. It is true I do not small

to deny that such a method of looking at it have not just as good a basis of reason, only, it appears to me at present that it can be defended just as little & certainty no better than the larval theory.

No one affirms that the intestinal diverticula which now in Phoxichilus & hymphon penetrate into the proboscis were originally [ancestrally] in the palps & ovigerous legs. The homology of the glandular organs of appendages II & III with parts of the sexual organs is not based on any observation ..... . But since I believe there are certain things which lessen Dohrn's arguments in favor of the Annelid theory, I believe that at bottom and the contents of pages 82 – 115 of his work confirms this opinion — the

ideas of this author & if mine upon the
Phylogeny as well for "Crustacea" as for
the Pycnogonids are not very different.
At first I saw only the weak side
of Dohrns Theory that he does not give
any explanation of the almost universal
presence in the ontogeny of these animals
of a characteristic larval form. (the larval
Protonymphon which for myself is a true
primary larva in the sense of Balfour
So that I cannot adopt a Theory which
causes to arise a form of Arthropods with
many segments from an Annelid with
the same number of appendages; but such
is not the opinion of Dohrn as clearly
expressed in the following paragraph.
Thus the larva of the greater part of the
Pycnogonids may be regarded, with a grain
of salt, as a form much like the ancestor.

" and if on the other hand. the absence of
an anal opening, the pincers of the first
pair of appendages, the long claws with
their accessory spines, the structure of
the proboscis with its triturating apparatus
& its ganglia, & the form of the cutaneous
glands with their integumentary hairs
can only be considered as having been
acquired in a much later stage & trans-
-mitted to larval life where they are
found at present; but which remains
in the larva that we may regard in reality
as inherited from its original condition.
Nothing but the nervous system ie the
the brain & ganglia, an intestine, three
pairs of appendages of a form modified
according to circumstances & two eyes!
But there are attributes which one
finds equally in the larva of Penaeus

with these segments. And if then we take into consideration that at one time the body of the Pycnogonids, as we believe, showed a great uniformity in the segments. (& the generative organs prove this) Then on the one hand there was a concentration & a differentiation & on the other hand a reduction in number of segments, these two conclusions lead us to admit a direct descent of the Pycnogonids from ancestral annelid-forms which were homonomously segmented

Then the larva of the Pantopods came from a larval form with pantopod-characters added on, but at the same time a larva which never had an independent & mature existence."

Whatever Hoek position may be with respect to The Pantopod-larva

with Dohrn in the most important part of the latters theory that the Pycnogonids have come down from the Annelids independently of the other groups of Arthropods. In examining the preceding account we may take it up, for the sake of clearness in two parts, the first pertaining to the ancestry of the adults leaving the larval form out of account & the second part where the meaning of the larval form will be considered.

I believe that if the account I have given of the early stages of development be even approximately correct there is little or no ground for a comparison between Crustacea & Pycnogonids certainly not with any existing forms.

The multipolar delamination of the

endoderm in the Pycnogonids has no homologue amongst the Crustacea nor is there any special similarity in the formation of the organs. There seems to be no trace of gastrulation in the ontogeny of the groups. And if we have reason for rejecting a relationship between the Pantopod-larva & the Nauplius — and I believe with Dohrn that we have — there remains nothing in common to the ontogeny of the two groups.

Nor are there any special affinities between the Annelids & Sea-Spiders, and there is one striking similarity between the latter & Peripatus, which I have already spoken of; but an isolated fact of this kind gives little ground for further comparison.

The neural organs in the two groups present a striking agreement but there is no proof forthcoming as to a real homology of the structures

The process of the formation of the endoderm described by Heider & by Wheeler in Insects shows a certain resemblance to multipolar delamination, but if it be such, in a more complicated form than shown by the Pycnogonids. With these two exceptions there seems to be nothing else in common in the ontogeny.

We are then left to decide between an independent origin for the Pycnogonids or a relationship with the Arachnids Prof Dohrn has ably maintained the first Theory & the preceding translation gave the conclusion he had reached.

Dr Hoek likewise, as we have seen, holds this opinion although not agreeing as to details with Prof Dohrn On the other hand a study of the early stages of the embryology has brought to light certain facts, which for me, point decidedly toward a community of descent between Arachnids & Pycnogonids. The latter show undoubted traces of degeneration & we cannot derive them from any existing animals. But I believe the Pycnogonids & the Arachnids have come down along the same line or in other words have had ancestors in common long after those ancestors came from Annelid-like forefathers.

The reasons for such a belief are as follows.

1st. The Pycnogonids form the endoderm
by a process of multipolar delamination
which is shown in its simplest form
in Phoxichilidium & Tanystylum &
in a more modified condition in
Pallene. In no other group of the
Triploblastica do we find a similar
phenomenon except in the Arachnids.
The description of the embryology
of Chelifer which Metchnikoff has
given shows this process, or something
quite analogous, to take place in it.
The segmentation is holoblastic & at
a later stage the large cells containing
yolk divide into an outer more
protoplasmic layer of cells & the
inner cells which are very granular.
The outer form the ectoblast & most
probably the inner, judging from his figure

for in the endoblast.

In the Spiders the process is not so well marked but if the conception which Balfour had of the formation of the yolk-nuclei be true, there we may make a direct comparison between the two groups. He says " It appears to me probable that at the time when the superficial layer of protoplasm is segmented off from the yolk below the nuclei undergo division & that a nucleus with surrounding protoplasm. is left with each yolk column" This description for the Spiders may be substituted word for word for the process of delamination of Pallene.

2nd. The first traces of the embryo to appear in Pallene is a round opaque area at the spot where the

stomodaeum invaginates.

In Schimkewitsch's recent account
for the Spiders he shows that the
primitive cumulus in them is the
place where the stomodaeum invaginates.
This is also true for Pallene but here
the stomodaeum invaginates quite early,
or perhaps simultaneously with the
early formation of mesoblast at this
place.   Further Schimkewitsch has
called attention to the fact that
the stomodaeum of Spiders in its earliest
developement is a triangular invagination & he compares it directly with
the triangular invagination of the
oesophagus of the Pycnogonids

3rd. The early formation of body cavity
surrounded by mesoblast in the leg
of Spiders has an exact parallel in

Pallene & Anoplodactylus. Yet
however tempting such a comparison
may be in this connection it must be
admitted I have not conclusively
proved this to be true for the Pycnogonids
but only exceedingly probable

4th In both Arachnids & Pycnogonids
we have well marked diverticula
from the mid-gut into the legs.

In the Pycnogonids these go into
the Chelicerae & the four pairs of
walking legs & the same holds for
the spiders but from a comparison of the
appendages of the two groups we must
suppose that the third pair of Pycnog
-onids appendages to have lost their
diverticula & the last appendages
either to have acquired diverticula or more probably
inherited them together with the appendage.

In Chelifer the diverticula appear very early in development & contain some of the yolk from the mid-gut. This is shown very distinctly in Metschnikoff's figures for Chelifer, & in this respect the embryo resembles closely the embryos of Pycnogonids.

5th In both Arachnids & Pycnogonids the first pair of appendages are chelate. This in itself would draw attention to the similarities of the two groups but we know further that in both groups this first pair of chelate appendages is innervated from the brain. These facts were considered by Balfour sufficiently important to indicate alone a relationship between the groups. He says.

" The presence of chelate appendages

innervated in the adult by the
supra oesophageal ganglia rather points
to a common Phylum for the Pycnogonida
& Arachnida, Though as shown above
all the appendages in the embryos of
true Arachnida are innervated by
post-oral ganglia."

I have not been able to find any
post-oral ganglia for Pallene but
the first pair of appendages arises
on the sides of the stomodaeum &
rather more forward. In this respect
it compares closely with the Pycnoda
& the early innervation of this pair
from the brain itself may be regarded
as a more abbreviated condition that
that was seen (by Balfour) in the Spiders
Metchnikoffs' figures for Chelifer
show the first pair of appendages

above on each side of the proboscis-like upper lip.

If future work verifies Dohrn (?)'s suggestion that this proboscis (rüsselförmige, provisorische Oberlippe) is homologous (entirely or in part) to the proboscis of the Pycnogonids — as his figure might indicate — then does the whole development of Chelifer show remarkably close resemblances to the Pycnogonids.

6th. The fourth pair of ambulatory legs — the seventh pair of appendages has been a stumbling block in the way of those who have compared Pycnogonid with Arachnid. Semper & Schimkewitsch have attempted to solve the difficulty by assuming that the third pair of appendages of the Pycnogonids — the ovigerous legs — are new structures of the

have called the four pairs of walking
legs homologous in the two groups.
Prof Dohrn has shown the impossibility
of dropping out in the count, the oviferous
legs & has shown that this pair of appendages are homonomous with the others.

The two pairs of Sense Organs in the
larval anterior ganglia we have
seen point unmistakeably to the same
conclusion, & give the final proof, if
such were really necessary.

We are led then to a comparison of
The appendages of the two groups beginning
with the chelicerae & going back pair for pair.
which leaves one pair over for the
Pycnogonids   Any comparison between
the two groups must Take into account this
extra pair of legs. Balfour has
suggested that this last argument is

appendages to represent the first abdominal
segment of the Arachnids.

The Third pair of appendages of the
Pycnogonids have assumed a special
function & at this time we might
suppose that an additional pair to
have been added on from the abdominal
segments. We also know that the embryos
of Spiders have rudimentary appendages
on the abdomen & as we have assumed
the Pycnogonids to have come from the
latter group most recently but remotely
when these appendages may have been
larger or even functional we have
some ground for a belief of such an
origin of the first segment.

There is another fact which may be
of importance in this connection
not only is the present pair of jointed _____

dorsalwards by the abdomen but at the same time the sixth pair of ganglia also is carried dorsalwards above the fifth pair, from which the connecting nerve fibres pass upwards to the sixth. At the present time it is impossible to determine whether this is due to a mechanical adjustment between the shifted abdomen & the last pair of thoracic ganglia or whether the sixth pair, belonging properly to the abdomen has taken part in the general shifting of that structure [*]

---

[*] If we assume with Hatchek a common descent for all arthropods, & that in the Insects we have several of the anterior segments, about the mouth, suppressed, we might assume that the Spiders have lost a third pair of appendages & the Pycnogonids retained it & in this way bring into line the other appendages of the groups.

There are certain objections against this comparison which I have attempted to see if they do not directly oppose yet do not support the hypothesis.

First & most important is the uncertainty of brain invaginations in Pycnogonids. These seem to be present in all the Arachnids & easily seen in the development of the embryo. In the Pycnogonids I have not been able to find such invaginations. I have seen in figures that the groove might possibly bear such an interpretation but even if this were true we would expect to find a much more pronounced involution but this does not seem to be the case. Again the Venhal Organs which have been compared with those of Peripatus lend no support to the hypothesis

It is possible that the Pycnogonids have come from the Arachnids at a time when the latter have had such organs in common with the ancestors of the Insects & that they have been fully retained in the Pycnogonids

Lastly the openings of the reproductive organs. Typically the ovaries & testes of the Sea-Spiders are a pair of organs united posteriorly by a cross commissure. They extend into the walking legs & open on the second joint of these. There are many exceptions to this but they are regarded as secondary.

The openings on the legs have no homologue in the Arachnids nor does it furnish any ground for comparison with other Arthropods. But if we assume the group to have come directly from the

Annelids we have no better ground
here for a comparison. We are greatly
in need of observations on the development of the sexual openings & until
we get such the question must remain
an open one.

It seems not improbable however
that the openings may be secondary
& connected in some way with the
secondary presence of reproductive
organs in the appendages

I hope to have shown, that these three
objections are of negative value, at
any rate so long as the present uncertain
-ty surrounds them, & that we have
sufficient grounds for a comparison
in the early stages of development, &
in some of the important adult
structures of the two groups.

The Pantopod-Larva.

There is a general resemblance between the Nauplius of the Crustacea & the Larva of the Pycnogonids, but the differences become greater & greater the more closely we examine the two forms. In each the body contains three pairs of appendages but those in the Nauplius show biramous structures while none of the Pantopod-Larval appendages show such structures. Moreover the first pair of appendages of the Pantopod-Larva is chelate & innervated from the brain. Other characteristics of the Larva are the well marked proboscis with its triangular oesophagus. The mid-gut sending out prolongations into some of the appendages & the absence of any anal opening.

Have we then any basis for the assumption
that this Pantopod-larva is a modified
Trochophore of the annelid ancestors?
The problem is very similar to that of
a supposed relationship between the
Nauplius & the Trochophore, & in both
cases an answer exceeding difficult
to give.   Prof Dohrn believes that
the Pantopod-larva is to be regarded
as a Trochophore with the Pycnogonid
characteristics reflected onto it.   Prof
Lang believes in a similar process
for the Nauplius   Without discussing
the latter let us confine our attention
to the possibilities of the Pantopod
larva.
What characters have been reflected
on Dohrns theory & what remains
after these are removed? Prof Dohrn

has given as the answer.

"   The absence of an anal opening,
The pincers of the first pair of appendages, The long claws with the accessory
spine, The structure of the proboscis
with its triturating-apparatus & its
ganglia, the form of the tegmentary
glands & their characteristic hairs,
can only be considered as having been
acquired in a much later stage, &
transmitted to larval life "
That is to say that almost every characteristic of the larva has been handed back from
the adult! And what remains?
" Nothing but the nervous system —
an intestine, three pairs of appendages
& two eyes. " " But there are the attributes
which one finds equally in the larva
of Annelids "! There is one last part

which has been left out of account, viz,
the presence of an anus in the
Trochophore. How can we account
for its absence on the above hypothesis
when we know the Pantopod-larva
to be a free living larval-form.
And unless some special reason for
such a loss can be imagined the
very basis of the comparison between
Pantopod-larva & Trochophore is gone!
What we have done in the above process
of subtraction is to have removed
the most striking structures of the adult
from the larva & have left — not a
Trochophore but only the frame work
of the Pycnogonid.
For two main reasons I am unable to
believe in the Phylogeny given by
Dohrn or by Hoek   First because

it seems to me there are facts derived from the early stages of development which point unmistakably to a relationship between Pycnogonids & Arachnids

And in the second place I cannot believe any actual homology to exist between the Protopod-Larva & the Trochophore, nor any fair reasons to assume that the characteristics of the Pycnogonids have been reflected into a Trochophore

If then we start with the assumption that there is a relationship between the Sea-Spiders & Arachnids we may examine into the meaning of the larval form as a corollary to such a position. I have stated that the Pycnogonids are degenerate & probably not directly derivable from

any existing groups. If so how far down in the ancestral tree of the Arachnids have they arisen The very great differences in the adult structure of the groups indicates no very recent origin and possibly they came on at a time when the Arachnids had the first pair of appendages chelate & there were innervated from the supraoesophageal ganglion; & had coeca from the digestive tract into most or all of the appendages. After the divergence of the Pycnogonids, as a group from the general phylum of the Arachnids the Pantopoda-larva may have developed

The Pycnogonids have adapted themselves to a very special habitat, so that there must be very great advantages for distribution & intercrossing in a free-

swimming independent larval form.
And in general what we may suppose to
have happened was a decrease in the
size & an increase in the number of eggs,
with the resulting early developement
of the free larva that we find to-day.
This larva represents the more anterior
segments of the adult viz, that part
containing the proboscis, the œsophagus
the eyes, the chelae, & two post-oral
appendages. Behind this is an undevel-
oped part which slowly grows in length
as the animal increases in size.
The anus belongs to the last segment
& does not appear until that segment
is wholly or in part developed.
If the Arachnids have come from Annelid
ancestors with many segments we have a clue
to the slight resemblance between the Pantopod

& The Trochophore. The former represents the most anterior segments of the adult Sea-Spiders & therefore to some extent the anterior segments of the Annelids or of the Trochophore. But at no time in the embryogeny of the Pycnogonids have Trochophore or Pantopod-larval characteristics existed separately as Dohrn believes.

Such seems to me the more probable view of the meaning of the Pantopod-larva. This belief has grown out of my work on the embryology of the group & whether future work supports or disproves such an hypothesis it is hoped it may be useful if only as furnishing another point of view for looking at the Phylogeny of the Pycnogonids, or may lead to a more complete study of the embryology of the group.

List of Works referred to.

1. Balfour.   Notes on the Development
              of the Araneina.
              Quart. Journ. Micro. Sc. Vol. XX. 1880.

2. Balfour. Elements of Embryology. 1880

3. Claus.   Crustaceen-Systeme.
                        Wien 1876.

4 Dohrn Ant.   Neue Untersuchungen
              über Pycnogoniden.
              Mittheil. a. d. Zool. Stat. Neapel
                        Band I 1879

5 Dohrn Ant. Die Pantopoden des
              Golfes von Neapel.
              Fauna und Flora des Golfes von Neapel
                  Band III   1881

6. Hatschek. Beiträge zur Entwicklungs-
     geschichte der Lepidopteren.
     Jena. Zeit. Naturwiss. Bd 11. 1877.

7 Heider. K. Ueber die Anlage der Keim-
     blätter von Hydrophilus piceus.
     Abh. d. K. Preuss. Akad d. Wiss.
                              Berlin 1885.

8 Müller, Fritz. Für Darwin.
                              Leipzig 1864.

9 Metschnikoff. E. Entwicklungsgeschichte
     des Chelifer.
     Zeit. f. wiss. Zool. XXI. 1871

10 Schimkewitsch. Zur Entwicklungsgeschichte
     der Araneen
     Zool. Anz. VII   1884

11. Schimkewitsch. Étude sur le développement des Araignées.
Archiv de Biologie 6 1887

12. Sedgwick. A. A monograph of the development of Peripatus Capensis.
Stud. Morph. Lab. Camb. IV. Pt. I 1885

13. Semper. Ueber Pycnogonida und ihre in Hydroiden schmarotzenden Larven-formen.
Arb. a. d. Zool. Zoot. Inst. Würzburg. 1. 1874.

14. Wheeler The embryology of Blatta Germanica & Doryphora Decemlineata.
Journ. Morph. III 1889.

15. Wilson. E. B. The Pycnogonida of New England & Adjacent Waters.
U. S. Commissioner Report Pt III 1878.

Reference Letters.
(for plates.)

B   = Brain.
D   = Mid-gut, mesenteron.
D' D" D'" D'" oc = Diverticula of midgut.
E   = Ectoblast.
N   = Entoblast.
M   = Mesoblast
P   = Periphery of Blastoderm.
S   = Schizocoeles
V' ⁻⁷ = Ventral Ganglia 1—7
H   = Heart
L   = Legs
O   = Opening Ventral Organs
C   = Central Cavity of Segmenting Egg.

Description of Plates.

Plate I

Fig 1. Section of segmented egg & membranes of Pallene empusa.

Fig 2. Section of another egg of Pallene at same stage to show central cavity C.

Fig 3. Section through upper pole of egg of Pallene. First formation of Blastoderm

Fig 4. Section through upper pole of egg of Pallene. Later stage than last Inner cells Entoblast are present.

Fig 5. Section through egg of Pallene. Later stage than last. Cap of cell — blastoderm - covers more of egg than in last.

Fig 6. Section through periphery of blastoderm at P of figure 5.

Plate 1. (continued)

Fig 7.   Section in same region as last

Fig 8.   Section of segmenting egg of Tanystylum
         orbiculare to show pyramidal shape of cells.

Fig 9. Section of egg of Tanystylum, to show
         process of multipolar delamination

Fig 10.   Same, with delaminated Ento blast

Fig 11.   Section of segmenting egg of Phoxi—
         —chilidium maxillare.

Fig 12.   Same, to show degeneration of
         endoderm.

Fig 13.   Cross-section of the Pantopod-larva
         of Tanystylum orbiculare.

Fig 14.   Circum-oesophageal ring of same

Fig 15.   Eyes of same.

Fig 16   Ventral Organs of same
         Early stage of third pair of
         ganglia

Plate II

Section of *Pallene empusa*.

Fig 17. Longitudinal section through
stomodaeum of young embryo.
First origin of mesoblast at M

Fig 18. Cross section through stomodaeum
of older stage. See line 18. Fig 1. Plate IV

Fig 19. Section anterior to last.
See line 19. Fig 1 Plate IV

Fig 20 Section anterior to figure 19
See line 20. Fig 1 Plate IV

Fig 21 Cross section of Ventral Organ.
See line 21. Fig 1. Plate IV

Fig 22 Cross section of Ventral Organs
at stage III

Fig 23. Cross section of Ventral Organs
at stage V

Fig 24 Cross section of Ventral Organs
at stage VI

Plate II (continued)

Fig 25.　Cross-section of embryo at
　　　　　Stage III

Fig 26.　Cross-section through brain &
　　　　　second pair ventral ganglia
　　　　　of Stage III

Fig 27.　Cross-section of embryo in the
　　　　　plane of a pair of legs.
　　　　　About Stage VI

Fig 28.　Cross-section of embryo in
　　　　　the plane between a pair
　　　　　of walking legs
　　　　　About Stage VI

Fig 28　Cross section through the
　　　　　rudimentary abdomen & the
　　　　　body of an embryo
　　　　　Stage VII -- VIII

Plate III.

A – J. Pallene empusa :
a – c. Phoxichilidium maxillare.
d – j. Tanystylum orbiculare.

Fig A   Segmenting egg, side view,
                    2-celled stage.

Fig B.   "   ", seen from above,
                    4-celled stage.

Fig C.   " seen from above,
                    8-celled stage.

Fig D.   " seen from below
                    8 celled stage.

Fig E.   "   " side-view
                    16 celled stage.

Fig F   "   " Surface views of
                    micromeres; 16 celled stage.

Fig G   "   " side view
                    16 macromeres, 8 micromeres.

Plate III (continued)

Fig H. Segmenting egg. side view.

24 macromeres

16 micromeres

Fig J Segmented egg. Lower Pole
Base of pyramidal cells
show at surface.

Fig. a.   Phoxichilidium 2-celled stage
Fig b.   "   4-   "   "
Fig C.   "   8   "

Fig d.   Tanystylum orbiculare 2-celled stage
Fig e.   "   "   4- "   "
Fig f.   32(?) "   "

Plate IX

1 – VIII   Pallene
IX   Tanystylum.

Fig I.   Surface view of young embryo, to
          show anterior region

Fig II   Surface view of ventral region
          of same.

Fig III   Ventral view of embryo, with
          four pairs ventral ganglia:
          first & second [?...]

Fig IV   Dorsal view of embryo about
          same age (a little older perhaps) as last.

Fig V   Ventral view of embryo. first
          & second ventral ganglia have fused.

Fig VI   Side view of embryo
          Showing rudimentary abdomen
          behind as a continuation of
          the body & in same plane as the body.

Plate IX (continued)

Fig VII   Ventral view of embryo at the
age when it leaves the parent
Fourth pair of walking legs
seen behind. Third pair of
appendages beginning just
in front of first pair of walking
legs.

Fig VIII   Dorsal view of last to show
stain & four eye spots
In both VII & VIII only the
proximal ends of the legs are
shown.

Fig IX   The Pantopod larva
(Protonymphon of Hoek) of
Tanystylum orbiculare.

Life

Thomas Hunt Morgan was born Sept 25th 1866. Preliminary education received in Public & Private Schools in Maryland. In '80 entered the Preparatory School of the State College of Kentucky; in the Autumn '82 entered the College & graduated in the Spring of '86. Entered Johns Hopkins University in the Autumn of '86. The Summer of '86 was spent at the Boston Marine Laboratory at Annisquam Mass; the Summer of '87 at the Harvard School of Botany: of '88 at Beaufort. N. C., & at the U.S Fish Commission Laboratory at Wood Holl, Mass; of '89 again at the last place.

The subjects offered for the degree of Doctor of Philosophy were Animal Morphology, Animal Physiology & Histology, & Morphological Botany.

www.ingramcontent.com/pod-product-compliance
Lightning Source LLC
Chambersburg PA
CBHW020350030726

47496CB00007B/2078